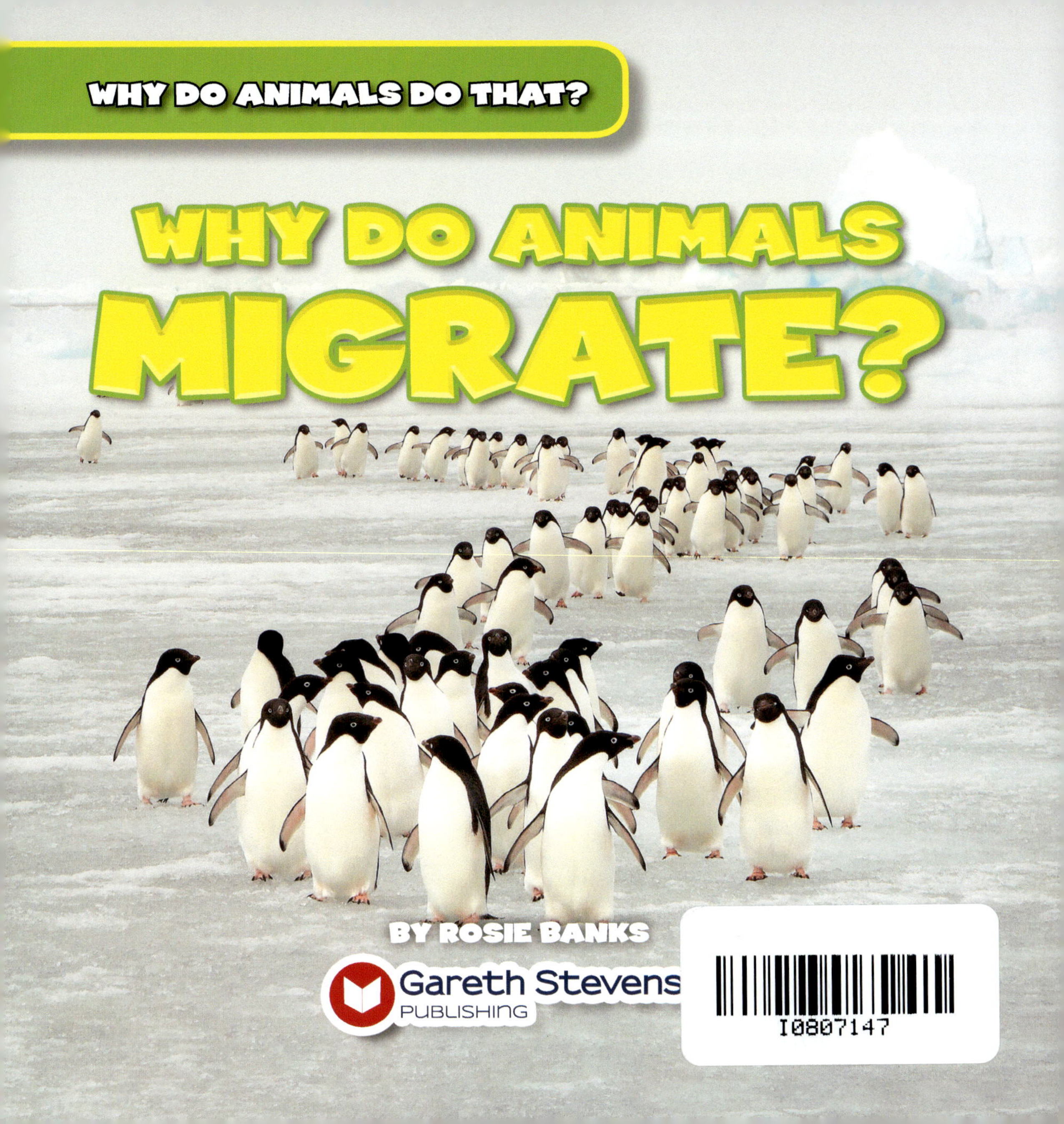

WHY DO ANIMALS DO THAT?

WHY DO ANIMALS MIGRATE?

BY ROSIE BANKS

Gareth Stevens PUBLISHING

Please visit our website, www.garethstevens.com. For a free color catalog of all our high-quality books, call toll free 1-800-542-2595 or fax 1-877-542-2596.

Library of Congress Cataloging-in-Publication Data

Names: Banks, Rosie, 1978- author.
Title: Why do animals migrate? / Rosie Banks.
Description: New York : Gareth Stevens Publishing, [2024] | Series: Why do animals do that? | Includes index.
Identifiers: LCCN 2022047777 (print) | LCCN 2022047778 (ebook) | ISBN 9781538285183 (library binding) | ISBN 9781538285176 (paperback) | ISBN 9781538285190 (ebook)
Subjects: LCSH: Animal migration–Juvenile literature.
Classification: LCC QL754 .B355 2024 (print) | LCC QL754 (ebook) | DDC 591.56/8–dc23/eng/20221018
LC record available at https://lccn.loc.gov/2022047777
LC ebook record available at https://lccn.loc.gov/2022047778

Published in 2024 by
Gareth Stevens Publishing
2544 Clinton Street
Buffalo, NY 14224

Designer: Claire Wrazin
Editor: Therese Shea

Photo credits: Cover Ivan Hoermann/Shutterstock.com; Series Art (texture) Anna Timoshenko, p. 5 Lipsett Photography Group/Shutterstock.com; p. 7 Karel Smilek/Shutterstock.com; p. 9 Sergey Uryadnikov/Shutterstock.com; p. 11 Olha Solodenko/Shutterstock.com; p. 13 Tomas Kotouc/Shutterstock.com; p. 15 Incredible Arctic/Shutterstock.com; p. 17 Dotted Yeti/Shutterstock.com; p. 19 sekarb/iStockphoto.com; p. 21 vladsilver/Shutterstock.com.

Printed in the United States of America

CPSIA compliance information: Batch #CSGS24: For further information contact Gareth Stevens, New York, New York at 1-800-542-2595.

CONTENTS

Migrating to Survive. 4
There and Back 6
Wildebeest by Land. 8
Whales by Water. 12
Arctic Terns by Air 14
The Monarch's Migration. 16
The Salmon's Return 18
Amazing Adaptation 20
Glossary. 22
For More Information. 23
Index 24

Boldface words appear in the glossary.

Migrating to Survive

Taking a trip to a far off place can be fun. But some animals need to make long trips each year just to **survive**! They might be looking for more food, a warmer **climate**, or a **mate**. This traveling is called migration.

There and Back

Many animals travel at some time in their lives. When some baby animals grow up, they find a new place to live. Migration is a different kind of movement. It means an animal goes somewhere and then returns to where it came from.

Wildebeest by Land

Many animals migrate to find food. More than a million **wildebeest** migrate each year. During the rainy season in southeastern Africa, wildebeest have grass to eat and water to drink. When the dry season begins, the grass and water dry up.

At the start of the dry season, wildebeest travel northeast, from Tanzania into Kenya. They cross rivers with crocodiles! They return to Tanzania when the rainy season begins. Some travel more than 500 miles (800 km) each year.

Whales by Water

Humpback whales have the longest ocean migration. In summer, they feed on fish in **polar** waters. In winter, they swim to warm waters to mate. They have their babies, called calves, in warm waters too. Some travel 5,000 miles (8,047 km) each year!

Arctic Terns by Air

Arctic terns have the longest migration! These birds **breed** in the Arctic in summer. Then they fly over 11,000 miles (17,703 km) to the Antarctic when it's summer there! The summer sunlight makes it easier for terns to find food.

The Monarch's Migration

The monarch butterfly is the only butterfly that migrates! It can't survive cold winters like other butterflies do. Some fly 3,000 miles (4,828 km) to reach warmer climates in Mexico and California. Sometimes, they catch a ride on air currents.

The Salmon's Return

Not every migration is yearly. Salmon **spawn** in rivers or lakes. After the baby salmon **hatch**, they swim to the ocean. Sometimes several years later, they swim to where they hatched. Then they spawn. They may jump over waterfalls on the way!

Amazing Adaptation

Many penguins migrate too. They travel far to breed and find food. You might wonder how animals know the way. Scientists think some sense Earth's **magnetic field** and use it like a map. Others might follow landmarks such as rivers. Some might use sunlight. There's a lot to learn about this **adaptation**!

WHY ANIMALS MIGRATE

To find a mate

To have babies

emperor penguins

To find food and water

To find a better climate

GLOSSARY

adaptation: A change in a type of animal that makes it better able to live in its surroundings.

breed: To mate and produce young.

climate: The average weather conditions of a place over a period of time.

hatch: To break open or come out of.

magnetic field: The area near a planet where magnetic forces can be found.

mate: One of two animals that come together to make babies. Also, to come together to make babies.

polar: Having to do with Earth's poles.

spawn: A way some marine animals reproduce in which eggs are let loose in the water.

survive: To live through something.

wildebeest: Another word for a gnu, an antelope that looks a bit like an ox with curved horns.

FOR MORE INFORMATION

BOOKS

Jaycox, Jaclyn. *Migration.* North Mankato, MN: Pebble, 2020.

Markle, Sandra. *Follow Those Zebras: Solving a Migration Mystery.* Minneapolis, MN: Millbrook Press, 2020.

Unwin, Mike. *Migration: Incredible Animal Journeys.* New York, NY: Bloomsbury, 2019.

WEBSITES

Animal Migrations
www.ducksters.com/animals/animal_migrations.php
Learn about more animals that migrate.

Monarch Butterfly
kids.nationalgeographic.com/animals/invertebrates/facts/monarch-butterfly
Read about dangers that this butterfly faces.

Publisher's note to educators and parents: Our editors have carefully reviewed these websites to ensure that they are suitable for students. Many websites change frequently, however, and we cannot guarantee that a site's future contents will continue to meet our high standards of quality and educational value. Be advised that students should be closely supervised whenever they access the internet.

INDEX

Africa, 8

Arctic terns, 14

babies, 6, 12, 18, 21

breed, 14, 20

California, 16

climate, 4, 16, 21

food, 4, 8, 12, 14, 20, 21

humpback whales, 12

Kenya, 10

mates, 4, 12, 21

Mexico, 16

monarch butterflies, 16

penguins, 20

salmon, 18

Tanzania, 10

water, 8, 12, 21

wildebeest, 8, 10